THE POETRY OF LITHIUM

The Poetry of Lithium

Walter the Educator™

SKB

Silent King Books a WhichHead imprint

Copyright © 2023 by Walter the Educator™

All rights reserved. No part of this book may be reproduced in any manner whatsoever without written permission except in the case of brief quotations embodied in critical articles and reviews.

First Printing, 2023

Disclaimer
This book is a literary work; poems are not about specific persons, locations, situations, and/or circumstances unless mentioned in a historical context. This book is for entertainment and informational purposes only. The author and publisher offer this information without warranties expressed or implied. No matter the grounds, neither the author nor the publisher will be accountable for any losses, injuries, or other damages caused by the reader's use of this book. The use of this book acknowledges an understanding and acceptance of this disclaimer.

"Earning a degree in chemistry changed my life!"
- Walter the Educator

dedicated to all the chemistry lovers, like myself, across the world

CONTENTS

Dedication v

Why I Created This Book? 1

One - Mind And Hand 2

Two - Science Imparts 4

Three - In All Its Forms 6

Four - Science And Art 8

Five - Lithium, Oh Lithium 10

Six - Science And Humanity 12

Seven - Source Of Elation 14

Eight - Peace To The Soul 16

Nine - Transforming Our World 18

Ten - Chemical Gem 20

Eleven - Great And Subtle 22

Twelve - Illuminating Life 23

Thirteen - World Of Science	. . . .	25
Fourteen - Force To Ignite	. . . .	27
Fifteen - Single Space	. . . .	29
Sixteen - Paradox So Grand	. . . .	31
Seventeen - Potent Dream	. . . .	33
Eighteen - Lightest Metal	. . . .	35
Nineteen - Healing Power	. . . .	36
Twenty - Smartphones To Laptops	. . . . .	38
Twenty-One - Mends The Seams	. . . .	40
Twenty-Two - Calming The Storms	. . . . .	42
Twenty-Three - Dusk Till Dawn	. . . .	44
Twenty-Four - Power And Peace	. . . .	46
Twenty-Five - Countless Wondrous Ways	. .	48
Twenty-Six - Element's Might	. . . .	50
Twenty-Seven - Unending Masterpiece	. . . .	52
Twenty-Eight - Plays Its Role	. . . .	54
Twenty-Nine - Sustainer Of Life	. . . .	56
Thirty - Intertwined Tight	. . . .	58
Thirty-One - Element Divine	. . . .	60
Thirty-Two - Progress And Change	. . . . .	62

Thirty-Three - Beauty Of Science	64
Thirty-Four - Whispers To Souls	66
Thirty-Five - Lithium's Touch	68
Thirty-Six - Electrons' Art	70
Thirty-Seven - Light Up Our World	72
About The Author	74

WHY I CREATED THIS BOOK?

Creating a poetry book about the chemistry element of Lithium was a unique and fascinating endeavor. Lithium, with its atomic number 3, has various interesting properties and applications, making it an intriguing subject for poetic exploration. By delving into the nature of Lithium, I can create poems that explore the element's atomic structure, its role in batteries and energy storage, its significance in medicine for mental health treatment, and its connection to the cosmos. This fusion of science and art can provide a fresh perspective, offering readers an opportunity to appreciate the beauty and complexity of both the scientific and poetic realms.

ONE

MIND AND HAND

In nature's realm, a radiant light,
Lithium, a marvel, shining so bright.
Within the periodic table's grace,
A symbol of power, in its rightful place.
   A metal of wonders, lithium's might,
Unleashing energy, igniting the night.
A beacon of hope, a fiery spark,
From deep within, where atoms embark.
   In stars' embrace, its genesis lies,
Forged in the furnace, where fusion defies.
Exploding supernovas, a celestial dance,
Lithium emerges, in cosmic expanse.
   On Earth's surface, it finds its home,
In rocks and minerals, where it roams.
A whisper of lithium, a silent call,
To scientists and dreamers, one and all.

In laboratories, its secrets unfold,
Unveiling its nature, both precious and bold.
Reacting with water, a sizzle and pop,
A vibrant reaction, a chemistry's top.

In batteries' core, lithium resides,
Endowing devices, with power that glides.
From smartphones to cars, a source divine,
Lithium's energy, a lifeline.

But lithium's tale holds more than just might,
It touches our souls, in the darkest of night.
A remedy for minds, unsteady and frail,
Calming the storm, a mental travail.

In medicine's grasp, its purpose is clear,
A mood stabilizer, diminishing fear.
A lifeline to those with minds astray,
Lithium brings solace, a guiding ray.

So, let us celebrate this element grand,
Lithium, a symbol of mind and hand.
From chemistry's realm to life's embrace,
Lithium's presence, a gift to embrace.

TWO

SCIENCE IMPARTS

In the realm of elements, bright and bold,
There shines a metal, precious and old.
Lithium, radiant and full of might,
A beacon of power, a dazzling light.

From the heart of a star, in a cosmic dance,
Lithium emerges, a product of chance.
In supernovas, where chaos reigns free,
This element forms, a celestial decree.

Earth, our home, cradles Lithium's embrace,
In rocks and in soil, it finds its place.
A silent guardian, hidden from view,
Yet its presence is felt, in all that we do.

In the realm of energy, Lithium shines,
A catalyst for progress, a force that defines.
In batteries it powers, our modern-day needs,
Unleashing potential, fulfilling our deeds.

But beyond science and technology's sway,
Lithium holds secrets, in a different way.
In the realm of emotions, it finds its use,
As a mood stabilizer, a balm to diffuse.

For those whose hearts are burdened with pain,
Lithium offers solace, a respite from strain.
A symbol of balance, of tranquility's grace,
It soothes the troubled minds, finding its place.

Lithium, a paradox, both powerful and meek,
In the periodic table, a treasure to seek.
A fusion of science and emotions, it blends,
An element that transcends, and mends.

Oh, Lithium, radiant and true,
We celebrate your essence, in all that you do.
From the depths of the cosmos, to the core of our hearts,
You're a symbol of wonder, where science imparts.

THREE

IN ALL ITS FORMS

In the realm of science, a paradox unfolds,
Lithium, the element, its story yet untold.
A metal of marvel, with secrets to confide,
A noble presence, in the world it does reside.

In laboratories, its essence is revered,
Unveiling mysteries, as its purpose is steered.
From batteries to fusion, it sparks innovation,
Powering our world with boundless fascination.

But beyond the confines of scientific lore,
Lithium's essence touches the depths of our core.
For in the human soul, emotions reside,
And Lithium's touch can heal and provide.

In medicine's embrace, it finds its role,
A tranquilizer for the mind, a soothing stroll.
Mending the fractures that anxiety brings,
Lithium's gentle touch mends broken wings.

In bipolar hearts, it brings balance anew,
A steady rhythm, where chaos once grew.
A stabilizing force, a conductor of peace,
Lithium's embrace, a sweet release.

Yet, Lithium dances in shades of gray,
A paradoxical element, it's safe to say.
For in its power lies a potential storm,
A double-edged sword, in its truest form.

But let us not forget, within duality's grasp,
Lithium's transformative power does clasp.
For in its nucleus, a fusion of science and soul,
A symphony of elements, a story to unfold.

So let us marvel at Lithium's might,
Both in science's realm, and emotions' light.
For in this paradox, a truth we shall find,
That Lithium, in all its forms, leaves no heart behind.

FOUR

SCIENCE AND ART

In the depths of celestial fires, it lies,
A whispering secret, a cosmic surprise.
Lithium, the spark that ignites the skies,
A symphony of atoms in harmonious ties.

From stellar nurseries, it was born,
Forged in the core of a star, it was sworn.
Into the universe, its essence was cast,
A wanderer of galaxies, moving so fast.

In batteries and power, it finds its worth,
A source of energy, a gift to the Earth.
From phones to cars, it powers our days,
Unleashing the potential in countless ways.

But beyond the realm of science and tech,
Lithium holds secrets we cannot neglect.
It touches the depths of our human soul,
A healer of emotions, making us whole.

The moods it can lift, the darkness it abates,
A remedy for hearts, a balm for the states.
In bipolar whispers, it dances with grace,
Balancing the mind, in its gentle embrace.

Oh, Lithium, element of duality,
From the cosmos to emotions, your versatility.
A symbol of wonder, both science and art,
You touch our world, healing every heart.

FIVE

LITHIUM, OH LITHIUM

In the realm of atoms, a star is born,
A metal of light, Lithium adorned.
With atomic number three, it takes its place,
A trace element that holds a cosmic embrace.

In the vastness of space, its genesis lies,
Forged in the fiery hearts of stars that rise.
From stellar explosions, it traveled afar,
To grace our world, a celestial memoir.

In the depths of Earth, its presence is found,
Embedded in rocks, deep underground.
Once hidden from sight, now it takes flight,
Unveiling its secrets, like a beacon of light.

In the realm of energy, Lithium thrives,
A catalyst for power that never dies.

From batteries to vehicles, it takes the lead,
Empowering our lives with its energetic creed.

    In the realm of medicine, it finds its way,
A healer of minds, a solace in gray.
For bipolar minds, it brings calm and peace,
Balancing emotions, granting sweet release.

    In the realm of emotions, it dances and weaves,
A conductor of feelings, where hearts find reprieve.
It softens the edges of sorrow and pain,
Infusing our souls with a gentle refrain.

    Lithium, oh Lithium, a paradox you are,
A metal of wonder, both near and far.
From the cosmos to our hearts, you bring your grace,
A symphony of elements, in perfect embrace.

SIX

SCIENCE AND HUMANITY

Lithium, oh Lithium, so light and so pure,
A metal that's rare, a treasure to secure.
Its atomic number is three, and it's a wonder to see,
How it reacts with water, creating a frenzy.
Lithium, oh Lithium, a mood stabilizer too,
A savior for many, a blessing oh so true.
It calms the mind and soothes the soul,
A chemical embrace that makes us whole.
Lithium, oh Lithium, a duality that's rare,
From energy to medicine, it's a wonder to share.
A miracle of science, a gift from above,
A symbol of transformation, that reminds us to love.
Lithium, oh Lithium, a healing power to behold,
A light in the darkness, a story to be told.

A shining star in the periodic table,
A versatile element that we should all be able
To appreciate and cherish, for all that it gives,
To science and humanity, it's a treasure that lives.

SEVEN

SOURCE OF ELATION

In the realm of science, a legend unfolds,
A metal so light, its story unfolds.
Lithium, the element, shining so bright,
It holds secrets that dance in the soft moonlight.

A spark of creation, a fiery birth,
In the hearts of the stars, it found its worth.
From the depths of the cosmos, it journeyed afar,
To the Earth's embrace, like a shining star.

In batteries, it powers our modern lives,
Fueling our devices and electric drives.
From smartphones to cars, it carries the load,
A silent companion on every road.

But Lithium's might goes beyond the mundane,
A healer of hearts, it eases the pain.
A guardian of minds, it brings harmony,
A mood stabilizer, a remedy.

    In pharmacies and hospitals, it finds its place,
Bridging the gaps between darkness and grace.
It soothes the storms that rage inside,
A beacon of hope, a calming tide.
    So let us celebrate this element divine,
A symbol of power, both yours and mine.
Lithium, you dance in the fires of creation,
A force for good, a source of elation.

EIGHT

PEACE TO THE SOUL

In the realm of elements, a star does shine,
A metal of light, both tender and fine.
Lithium, the atom, with atomic number three,
Unveiling secrets with its chemistry.

A paradox it holds, this element so pure,
Dual in nature, both stable and obscure.
Its electrons dance, a vibrant energy,
Igniting fires of curiosity.

In laboratories, scientists explore,
Unlocking the mysteries at its core.
From batteries to medicine, it finds its place,
Transforming lives with its healing grace.

In the realm of emotions, lithium's touch,
A soothing balm, it offers so much.
It calms the storms within, brings peace to the soul,
A chemical solace, making broken hearts whole.

Yet lithium's power extends beyond,
To the realm of energy, where it's fond.
In batteries it resides, a vital force,
Empowering devices with its resource.

So let us celebrate this element of might,
Lithium, the catalyst, shining so bright.
From science to emotions, it weaves its spell,
A transformative power we know so well.

NINE

TRANSFORMING OUR WORLD

In the realm of elements, let us delve,
To a metal that holds stories untold, they tell,
Lithium, a paradox of secrets it keeps,
A dance between science and emotions it weaves.

In laboratories, it's a scientist's delight,
With atomic number three, shining so bright,
Its electrons, they buzz, in a dance so wild,
A conductor of energy, a catalyst of life.

But beyond the lab, where feelings reside,
Lithium whispers, a healer inside,
It dances through neurons, calming the storm,
A balm for the mind, a tranquility reborn.

Bipolar emotions, they find solace in its touch,
Balancing the pendulum, it means so much,

A stabilizer of moods, a guardian of peace,
Lithium's embrace, a sweet release.

In the realm of energy, it holds a key,
Powering our devices, setting us free,
Batteries charged with its electric might,
Lithium's energy, illuminating our night.

And in medicine's realm, it finds its place,
A remedy for ailments, a gift of grace,
From the depths of depression to the highs of mania,
Lithium's touch, a divine panacea.

Oh, Lithium, a paradox so profound,
A catalyst, a healer, in science renowned,
From the lab to the mind, its wonders unfurled,
Lithium, the element, transforming our world.

TEN

CHEMICAL GEM

In the realm of atoms, Lithium stands,
A chemical gem, its power expands.
Electron dance, with three it begins,
A nucleus strong, where its essence resides within.

Lithium, the element of vibrant flame,
Its light shines bright, a celestial acclaim.
From stars it's born, in cosmic fires,
A gift to Earth, with magnetic desires.

In batteries, its energy unfurls,
Powering our lives, as it unfurls.
Electric dreams, electric cars,
Lithium's spark, a world that soars.

But Lithium's touch extends beyond,
To minds in need, to hearts that long.
In medicine's hands, a healing embrace,
A balm for souls, a gentle grace.

    Emotions tamed, bipolar's hold,
Lithium's whispers, a story told.
Odes to its power, a poet's voice,
A bridge between worlds, a choice.
    So let us marvel at Lithium's might,
In science and art, a guiding light.
From energy's surge to emotions' sway,
Lithium, a muse, forever will stay.

ELEVEN

GREAT AND SUBTLE

Lithium, a metal so light,
It powers our devices with might,
From phones to cars, it's the key,
To keep them running smoothly and free.

But beyond its electronic role,
Lithium has the power to console,
In medicine, it's used to treat,
Bipolar disorder, a feat.

This element, so versatile,
Has a presence in our lifestyle,
From the practical to the healing,
Lithium's properties are revealing.

So let us celebrate this metal,
Its impact both great and subtle,
For Lithium, in all its forms,
Is a vital part of our norms.

TWELVE

ILLUMINATING LIFE

In the realm of elements, Lithium shines,
A treasure of science, where wonders align.
With its atomic number of three,
It brings forth a world of possibility.

In the realm of power, Lithium reigns supreme,
Powering devices, a creator's dream.
From smartphones to laptops, it fuels our needs,
A source of energy, where progress succeeds.

In the realm of healing, Lithium's embrace,
A remedy for souls, in tranquility and grace.
With its calming touch, it soothes the mind,
A refuge for emotions, where solace we find.

In the realm of wonder, Lithium unfolds,
Unraveling secrets, as the story unfolds.
From the stars above to the depths of the Earth,
Lithium's presence, a testament of its worth.

So let us celebrate this element divine,
Lithium, a gift, so rare and fine.
In science, in medicine, in everyday strife,
Lithium, a beacon, illuminating life.

THIRTEEN

WORLD OF SCIENCE

In the realm of chemistry, Lithium stands tall,
A wondrous element, with powers enthrall.
From batteries to medicine, its uses are vast,
A substance that impacts our present and past.

In devices it thrives, providing energy and might,
Powering our gadgets, shining ever so bright.
From smartphones to laptops, it plays a key role,
Lithium's power, an integral part of the whole.

But beyond the realm of technology's hold,
Lithium's healing properties begin to unfold.
In psychiatry, it finds its profound place,
Calming the restless, soothing the mind's race.

For those with ailments that grip the soul,
Lithium brings solace, helping them feel whole.
A remedy for mood swings and bipolar strife,
Lithium brings balance, restoring inner life.

So let us celebrate this element divine,
For its power and healing, like a sacred shrine.
Lithium, a marvel in the world of science,
A force that shapes our lives with its benevolence.

FOURTEEN

FORCE TO IGNITE

In the realm of power and light,
There exists a metal, shining so bright,
Lithium, the element, a force to ignite,
A marvel of chemistry, delicate and slight.

Within the batteries that power our days,
Lithium dances, in its own mysterious ways,
Unleashing energy, like a solar blaze,
It fuels our devices in countless arrays.

But beyond the realm of technology's might,
Lithium holds secrets, hidden from sight,
For in the field of medicine, it takes flight,
A healer, a soother, bringing respite.

In minds troubled by darkness and despair,
Lithium whispers hope, a gentle repair,
It calms the storms that rage and tear,
Restoring balance, banishing the nightmare.

So let us raise a toast to Lithium's grace,
In power, in healing, it finds its rightful place,
A paradox, a wonder, with a mystic embrace,
Lithium, the element, we forever embrace.

FIFTEEN

SINGLE SPACE

In the realm of chemistry, a star does shine,
With properties unique, a gem so fine.
Lithium, the element, its secrets unfold,
A power to behold, a story yet untold.

In the realm of devices, Lithium takes its place,
Powering our gadgets with its electric embrace.
From smartphones to laptops, it fuels our delight,
A catalyst of progress, shining so bright.

But beyond the circuits, an enigma it hides,
In the realms of medicine, where healing resides.
A balm for the mind, an elixir of calm,
Lithium, the healer, with its soothing charm.

Mending the fractures of a restless soul,
Balancing the chaos, making us whole.
A remedy for the weary, a salve for the pain,
Lithium, the savior, brings peace in its reign.

So let us marvel at this element divine,
As it powers our world and heals our mind.
Lithium, the enigma, with its dual grace,
A force of nature, in every single space.

SIXTEEN

PARADOX SO GRAND

In the realm of chemistry, let us explore
The wonders of an element, Lithium we adore
From the depths of the periodic table it springs
A metal so tranquil, yet it carries many things
In the world of power, it finds its humble place
Batteries and devices it helps to embrace
With electrons dancing, it charges them all
Lithium-ion, a force that never shall fall
But beyond the realm of technology's might
Lithium's healing touch shines bright
In medicine, it finds its sacred role
A balm for troubled minds, it consoles
From bipolar disorder to depression's toll
Lithium's embrace brings solace to the soul
A stabilizer, a guardian of harmony
Restoring balance, setting spirits free

Oh, Lithium, you are a paradox so grand
A metal of power, a healer's helping hand
Through science and medicine, your worth is revealed
A versatile element, in awe we are sealed.

SEVENTEEN

POTENT DREAM

Lithium, oh element divine,
In the realm of atoms, you truly shine.
A power stored within your core,
Unleashed in devices, forevermore.

In the world of technology, you thrive,
Beneath screens and circuits, you come alive.
From batteries small to electric cars,
You power our lives, like shooting stars.

But beyond the gadgets, let me tell,
Of your secrets, Lithium, I shall unveil.
In medicine, you bring healing light,
A balm for minds lost in the night.

For those who struggle with deep despair,
You bring a calmness, a tender care.
Mood stabilizer, a guiding hand,
Lifting the burden, helping them stand.

Lithium, a paradox you may seem,
In science and health, a potent dream.
From charging devices to soothing souls,
Your presence, Lithium, truly consoles.

EIGHTEEN

LIGHTEST METAL

In the realm of elements, one stands apart,
With properties unique, it plays its part,
Lithium, the lightest metal of them all,
From the heart of stars, it did once fall.

In technology's embrace, it finds its place,
Powering devices with its electric grace,
In batteries and smartphones, it sparks the fire,
Fueling our desires, taking us higher.

But Lithium's story goes beyond machines,
In medicine, its healing touch is seen,
A mood stabilizer, a calmness it imparts,
Easing troubled minds, mending broken hearts.

From the depths of the Earth, Lithium does rise,
A symbol of hope, a gift in disguise,
With power to heal, to bring balance and peace,
Lithium, the element, our souls release.

NINETEEN

HEALING POWER

In the realm of elements, a star does shine,
Lithium, an element of power divine.
With three protons, it holds a special place,
In the cosmos, it dances with grace.

A metal so light, it floats on water's crest,
Its presence in our lives, it truly does attest.
In technology's embrace, it finds its worth,
Powering our devices, giving them birth.

From batteries small to electric cars,
Lithium's energy, it truly stars.
It charges our lives, keeps us connected,
A force of innovation, never neglected.

But Lithium's magic extends beyond machines,
In medicine, its healing power convenes.
A remedy for minds that tremble and ache,
Calming the storms, bringing solace in its wake.

A mood stabilizer, a guardian of the mind,
Lithium's touch, a tranquilizer we find.
For bipolar souls, it brings a steady hand,
Balancing emotions, like a maestro's band.

So let us celebrate this element true,
Lithium, a treasure with much to pursue.
From technology's marvels to healing's embrace,
Lithium's presence, a gift we embrace.

TWENTY

SMARTPHONES TO LAPTOPS

In the realm of science's embrace,
Lithium, a noble element, takes its place.
A catalyst for life, a spark in the dark,
It powers our devices, a technological arc.

In batteries it resides, a source of might,
Unleashing energy, like a beacon's light.
From smartphones to laptops, it brings us ease,
Connecting the world, with effortless breeze.

But beyond the realm of circuits and wires,
Lithium's healing touch never tires.
In medicine, its power is revealed,
A soothing balm for wounds concealed.

It calms the mind, a tranquil repose,
Restoring balance, relieving woes.

An ally for those who suffer within,
Lithium's healing touch, a gentle spin.
    So let us celebrate this element divine,
A paradoxical force, both fierce and benign.
From powering our world with technological grace,
To healing our souls, with a tranquil embrace.

TWENTY-ONE

MENDS THE SEAMS

In the realm of science, a humble light,
Lithium, an element, shining so bright.
A metal of wonder, a spark of might,
In the depths of chemistry, its secrets ignite.

From the stars it emerged, a cosmic birth,
A gift to Earth, a treasure of worth.
In technology's embrace, it finds its place,
Powering our devices, with energetic grace.

In batteries it resides, a pulsating force,
Electric currents flow, a relentless source.
From smartphones to electric cars,
Lithium's power takes us far.

But beyond the realm of circuits and wires,
Lithium's healing touch never tires.
In medicine's domain, its role is grand,
A soothing balm for minds that demand.

For those who bear the weight of despair,
Lithium's embrace offers solace and care.
It calms the storms that rage within,
Restoring balance, where chaos has been.

So let us marvel at Lithium's might,
A versatile element, shining so bright.
In technology's realm, it fuels our dreams,
While in healing's embrace, it mends the seams.

TWENTY-TWO

CALMING THE STORMS

In the realm of elements, Lithium reigns,
A spark of wonder, where power sustains.
From the depths of stars, it came to be,
A gift from nature, for all to see.

In technology's grasp, it finds its place,
A conductor of energy, with boundless grace.
In batteries and capacitors, it resides,
Empowering our devices, as it abides.

Yet beyond the circuits and wires it fuels,
Lithium holds secrets, profound and true.
In medicine's realm, it finds its worth,
A healer of minds, a balm for the earth.

For troubled souls seeking solace and peace,
Lithium's touch brings a gentle release.

Calming the storms that rage within,
Restoring balance, where chaos has been.
　Oh, Lithium, dual natured and grand,
In technology's realm, you firmly stand.
But it is in your healing touch we find,
A remedy for hearts and troubled minds.
　So let us celebrate this element divine,
For its power in technology, and peace it defines.
Lithium, a paradox, both fierce and serene,
A marvel of science, a force to be seen.

TWENTY-THREE

DUSK TILL DAWN

In realms of science, where mysteries unfold,
There lies a secret, a treasure untold.
Lithium, the element of wonders profound,
In nature's realm, its essence is found.

A spark of power, it holds within,
Unleashing energy, where new worlds begin.
In batteries, it dances with electrons,
Powering devices, from dusk till dawn.

But beyond the realm of technology's might,
Lithium extends its healing light.
A guardian of minds, it finds its place,
In medicine's embrace, a saving grace.

For troubled souls, it brings peace and calm,
Dispelling darkness, like a soothing balm.
In bipolar battles, it finds its call,
Balancing emotions, preventing the fall.

    Oh, lithium, a paradox you possess,
A force of nature, a gentle caress.
From circuits to minds, you lend your might,
A symphony of elements, harmonious and bright.
    So let us marvel at your dual grace,
In chemistry's realm, a sacred space.
Lithium, the element that shines so bright,
In technology's realm, and hearts' darkest nights.

TWENTY-FOUR

POWER AND PEACE

In circuits and cells, Lithium abounds,
A dual-natured element, marvel renowned.
In the realm of technology, it takes the lead,
Powering devices with remarkable speed.

From batteries small to electric cars grand,
Lithium's energy fuels progress by hand.
It conducts the spark that ignites innovation,
Pushing boundaries with boundless fascination.

But beyond the realm of wires and volts,
Lithium's healing touch brings solace and hope.
In psychiatric realms, it finds its place,
Calming troubled minds, a tranquil embrace.

A remedy for the restless, a balm for the soul,
Lithium whispers peace, making broken hearts whole.
Its gentle touch brings balance and calm,
A remedy hailed, a healing balm.

So let us celebrate this paradoxical force,
That shines bright in realms diverse, of course.
Lithium, the element of power and peace,
Bringing harmony to both science and release.

TWENTY-FIVE

COUNTLESS WONDROUS WAYS

In the realm of atoms, a wondrous tale unfolds,
Of an element, Lithium, with secrets yet untold.
A paradox it carries, a duality profound,
Both a source of energy and a healing compound.

In the realm of technology, it sparks with fiery might,
Powering our devices, like stars in the night.
Batteries charged with potential, they hum and they gleam,
Lithium, the catalyst for our digital dream.

But beyond the circuits and wires, in the realm of the mind,
Lithium casts its gentle touch, a solace we find.
A soothing balm for troubled souls, a remedy so rare,

It whispers calm to chaos, and mends hearts in despair.
Like a steady hand in darkness, it guides us through the storm,
Easing the weight of burdens, bringing peace in any form.
In psychiatric realms, its touch is gentle and serene,
Lithium, the healer, the quiet antidote to the unseen.
A paradox it carries, this element of might,
Fueling progress and innovation, while healing in the night.
Lithium, oh Lithium, a force both fierce and kind,
Whispering peace to troubled minds, making broken hearts whole.
So let us celebrate this element, a true marvel of our days,
A potion of power and peace, in countless wondrous ways.
In science and release, it brings harmony's sweet embrace,
Lithium, the element that fills our world with grace.

TWENTY-SIX

ELEMENT'S MIGHT

In the realm of atoms, a dual force resides,
Lithium, the element that gracefully glides.
A paradox it holds, in its atomic core,
Power and healing, a tale to explore.

In the world of technology, Lithium reigns,
A source of energy, it proudly sustains.
Batteries and power cells, it does empower,
Fueling our gadgets, every passing hour.

Yet beyond the circuits, a secret it keeps,
A healer of minds, a solace it weeps.
In tiny white tablets, it brings relief,
Calming the storms, easing the grief.

Lithium, the paradox, a marvel indeed,
Harnessing energy, fulfilling our need.
But also a balm, a tranquilizer of souls,
Restoring the balance, making us whole.

So let us embrace this element's might,
In technology's realm, in the darkest of night.
For Lithium, the paradox, a force so rare,
Brings power and healing, a cosmic affair.

TWENTY-SEVEN

UNENDING MASTERPIECE

In the realm of chemistry, behold the might,
A metal so potent, a wondrous sight.
Lithium, the element of pure energy,
Unleashing power, with a fiery decree.

From the depths of the Earth, it does arise,
Igniting innovation, reaching the skies.
A spark of invention, a catalyst profound,
Lithium, the force that propels us, unbound.

In batteries it dwells, a source of might,
Empowering devices, day and night.
From phones to cars, it fuels our desires,
Lithium, the driver of technological fires.

But beyond the realm of circuits and wires,
Lithium transcends, its healing never tires.

For in medicine's grasp, it finds its place,
A balm for the soul, a solace in grace.

A tranquilizer, calming the storm within,
Lithium, a salve for minds that have been.
With delicate balance, it brings us peace,
A gentle touch, a longing release.

Like a guardian angel, it watches over the mind,
Restoring harmony, leaving no pain behind.
Lithium, the paradox, both fierce and serene,
A force of nature, a healer unseen.

So let us celebrate this element of might,
That fuels progress and brings solace to the fight.
Lithium, the essence of power and peace,
A testament to nature's unending masterpiece.

TWENTY-EIGHT

PLAYS ITS ROLE

In the realm of atoms, a force so bright,
Lithium emerges, a beacon of light.
Its electrons dance, a delicate ballet,
With power and peace, they lead the way.

In the realm of technology, it finds its place,
A catalyst for progress, a boundless embrace.
Batteries humming, devices alive,
Lithium's energy, ready to thrive.

But beyond the circuits and screens that glow,
Lithium holds secrets, few others know.
In the realm of minds, it's a healer profound,
Calming the chaos, bringing solace around.

For troubled souls, it offers respite,
A balm for the weary, a soothing light.
Depths of despair, it gently lifts,
A flicker of hope, a spirit that shifts.

Lithium, a paradox, both fire and balm,
A bridge between worlds, a soothing calm.
In chemistry's realm, it plays its role,
A force of nature, both powerful and whole.

TWENTY-NINE

SUSTAINER OF LIFE

In the realm of atoms, a power so bright,
Lithium emerges, a radiant light.
A paradox it holds, both fire and grace,
A source of energy, a healer's embrace.

From fiery cores of stars, its journey begins,
A cosmic gift, where creation begins.
With three electrons, it dances and spins,
A catalyst of life, where hope never dims.

In the realm of hearts, a solace it brings,
Lithium, the balm for troubled minds.
An elixir of calm, a gentle caress,
It soothes the storms, brings peace, no less.

A dual nature it holds, both fierce and serene,
Lithium, the enigma, a paradoxical dream.
In the hands of progress, it fuels innovation,
In the hearts of souls, it brings restoration.

So let us marvel at this element's might,
Lithium, the fusion of darkness and light.
A force that empowers, a sustainer of life,
In energy and healing, forever it thrives.

THIRTY

INTERTWINED TIGHT

In the realm of elements, a paradox is found,
A metal so light, yet with power unbound.
Lithium, they call it, a wondrous creation,
A source of energy and mind's reparation.

From the depths of the Earth, it does emerge,
A treasure to harness, a power to surge.
In batteries, it dances, electrons aglow,
Powering our devices, from high to low.

But Lithium's tale is not limited to tech,
It holds a secret, a healing effect.
In psychiatric realms, it finds its place,
A balm for troubled minds, a saving grace.

It calms the storms within, the raging tides,
Easing the burden that the soul abides.
A stabilizer of moods, a tranquilizer too,
Lithium, the healer, bringing peace anew.

So, behold the paradox, both power and peace,
Lithium's duality, never to cease.
A force of nature, a chemist's delight,
Innovation and solace, intertwined tight.

THIRTY-ONE

ELEMENT DIVINE

In the realm of atoms, a jewel is found,
A metal so light, yet profound,
Lithium, the element, both fierce and meek,
A paradoxical nature, it does speak.

In the depths of the earth, its secrets concealed,
A power untapped, soon to be revealed,
Reactive and fiery, it dances with flame,
Igniting the world with its shimmering fame.

But beyond its surface, a deeper tale,
Lithium's essence, a soothing gale,
For troubled minds, it brings a calming grace,
A balm for the soul, a gentle embrace.

In batteries, it powers the modern age,
Electricity flows, the world's new stage,
From smartphones to cars, it fuels our dreams,
Lithium, the catalyst, bursting at the seams.

Yet in medicine's realm, it finds its place,
An elixir of peace, a saving grace,
For bipolar hearts, it steadies the storm,
Lithium, the healer, a refuge warm.

So let us behold this element divine,
A dichotomy of power and peace entwined,
Lithium, the enigma, a paradox untold,
In chemistry's tapestry, a wonder to behold.

THIRTY-TWO

PROGRESS AND CHANGE

In the realm of healing, Lithium shines,
A potent element, both gentle and divine.
A guardian of minds, it holds the key,
Unlocking serenity, setting spirits free.

In medicine's hands, its power unfolds,
A remedy for anguish, where hope takes hold.
Bipolar's tumultuous storms it can subdue,
Balancing emotions, bringing peace anew.

A chemist's delight, a catalyst supreme,
Lithium dances, igniting dreams.
From fiery reactions, creations arise,
In labs of discovery, it mesmerizes.

Yet Lithium's duality is what sets it apart,
A paradoxical element, a balm for the heart.

From chemistry's crucible to minds' sanctuary,
Lithium's magic weaves a tale contrary.
    Its atomic dance fuels progress and change,
While soothing the restless, a tranquil exchange.
With each electron's dance, a symphony unfolds,
Both catalyst and solace, as time unfolds.
    Lithium, a paradox in its core,
Mending the broken, mending the sore.
A healing elixir, both subtle and grand,
In its gentle touch, tranquility expands.
    So let us celebrate this element pure,
In chemistry's embrace, and minds' allure.
For Lithium's enchantment, its powers untold,
Bring harmony and solace, to young and old.

THIRTY-THREE

BEAUTY OF SCIENCE

In the realm of elements, there lies a gem,
A paradoxical force, Lithium, its name.
A catalyst for progress, a balm for the heart,
A substance that sets worlds alight, yet soothes the flames.

With atomic number three, it claims its place,
A metal light as air, a chemist's delight.
In batteries it resides, a powerhouse unseen,
Empowering technology, in day and night.

But beyond the realm of wires and circuits,
Lithium's magic extends to troubled minds.
A healer, a calmer, it holds a special key,
Unraveling the storms, where peace it finds.

Oh Lithium, you are a paradoxical wonder,
A force of nature, both fierce and serene.

Innovation and restoration, hand in hand,
In your presence, marvels are foreseen.
   So let us raise a toast to this element grand,
To Lithium, the catalyst and the balm.
For in its dance of contradictions, we find,
The beauty of science, a lyrical charm.

THIRTY-FOUR

WHISPERS TO SOULS

In a realm of elements, Lithium resides,
A dual nature it holds, where opposites collide.
A force in technology, powering our devices,
Yet a healing element for troubled minds' vices.

In the depths of its structure, electrons dance,
A vibrant energy, in a cosmic trance.
A metal so light, it floats on liquid's embrace,
A symbol of innovation, pushing boundaries with grace.

In lithium-ion batteries, it finds its home,
Powering our world, wherever we may roam.
In electric cars, it fuels the drive,
A sustainable future, for which we strive.

But beyond the realms of wires and circuits,
Lithium whispers to souls, its healing virtues.

A balm for troubled minds, it brings tranquility,
Restoring balance, with gentle serenity.
    In the hands of healers, it finds its place,
Treating the scars of the human race.
A catalyst for stability, a calming force,
Lithium brings solace, its power a discourse.
    So let us marvel at Lithium's might,
Both a force of progress and a healer's light.
In technology and minds, it intertwines,
A paradoxical element, where beauty shines.

THIRTY-FIVE

LITHIUM'S TOUCH

In the realm of chemistry's might,
Lithium shines, a radiant light.
A paradox it holds within,
A dual nature, a dance begin.

    In the depths of batteries it resides,
Powering our world with quiet strides.
Electrons flow, a current strong,
Lithium's force, it drives us along.

    But beyond the circuits and wires,
Lithium's touch, it inspires.
A balm for the restless minds,
A salve for the troubled finds.

    Its ions dance, in medicines pure,
Restoring balance, a tranquil cure.

For those who wander in the dark,
Lithium's embrace, a soothing spark.
    So let us marvel at its grace,
A fusion of power and solace's embrace.
Lithium, a paradox, a force untamed,
Innovation's fire, healing's flame.

THIRTY-SIX

ELECTRONS' ART

In the realm of elements, a paradox unfolds,
Lithium, a force both mighty and bold.
With three electrons, it dances with grace,
A symphony of atoms, an ethereal embrace.

In the realm of technology, it reigns supreme,
Powering our devices, a futuristic dream.
Batteries charged with its electric might,
Igniting progress, a beacon of light.

But beyond the circuits and wires it fuels,
Lithium holds a secret, a balm for troubled souls.
In the realm of healing, it finds its way,
Calming storms within, bringing peace to stay.

A salve for the restless mind, it gently weaves,
Easing the burden, granting tranquility's reprieve.
Lithium, a paradox, both fire and balm,
Harnessing energy, while soothing with calm.

So let us celebrate this enigmatic element,
A catalyst for progress, a healer's testament.
Lithium, the dancer of electrons' art,
A force that shapes both science and heart.

THIRTY-SEVEN

LIGHT UP OUR WORLD

Lithium, oh Lithium, a metal so rare,
Found in the earth, but also in air.
A paradox you are, both strong and calm,
A force of innovation, a healer's balm.

Your electrons dance, a wild frenzy,
Powering batteries, a modern-day frenzy.
But also a remedy for troubled minds,
A soothing touch, a rare find.

You light up our world, oh Lithium bright,
A catalyst for progress, a shining light.
Yet in a pill, you bring peace to the troubled,
A calming force, a mind untroubled.

Lithium, oh Lithium, a metal so rare,
A dual nature, a paradox so fair.

Power and healing, together you bring,
A precious element, a rare thing.

ABOUT THE AUTHOR

Walter the Educator is one of the pseudonyms for Walter Anderson. Formally educated in Chemistry, Business, and Education, he is an educator, an author, a diverse entrepreneur, and he is the son of a disabled war veteran. "Walter the Educator" shares his time between educating and creating. He holds interests and owns several creative projects that entertain, enlighten, enhance, and educate, hoping to inspire and motivate you.

Follow, find new works, and stay up to date
with Walter the Educator™
at WaltertheEducator.com

www.ingramcontent.com/pod-product-compliance
Lightning Source LLC
LaVergne TN
LVHW020133080526
838201LV00117B/3716